THEORETICAL ANALYSIS OF HIGH FUEL UTILIZING SOLID OXIDE FUEL CELLS

THEORETICAL ANALYSIS OF HIGH FUEL UTILIZING SOLID OXIDE FUEL CELLS

PEDRO NEHTER

Nova Science Publishers, Inc.
New York

Copyright © 2009 by Nova Science Publishers, Inc.

All rights reserved. No part of this book may be reproduced, stored in a retrieval system or transmitted in any form or by any means: electronic, electrostatic, magnetic, tape, mechanical photocopying, recording or otherwise without the written permission of the Publisher.

For permission to use material from this book please contact us:
Telephone 631-231-7269; Fax 631-231-8175
Web Site: http://www.novapublishers.com

NOTICE TO THE READER

The Publisher has taken reasonable care in the preparation of this book, but makes no expressed or implied warranty of any kind and assumes no responsibility for any errors or omissions. No liability is assumed for incidental or consequential damages in connection with or arising out of information contained in this book. The Publisher shall not be liable for any special, consequential, or exemplary damages resulting, in whole or in part, from the readers' use of, or reliance upon, this material.

Independent verification should be sought for any data, advice or recommendations contained in this book. In addition, no responsibility is assumed by the publisher for any injury and/or damage to persons or property arising from any methods, products, instructions, ideas or otherwise contained in this publication.

This publication is designed to provide accurate and authoritative information with regard to the subject matter covered herein. It is sold with the clear understanding that the Publisher is not engaged in rendering legal or any other professional services. If legal or any other expert assistance is required, the services of a competent person should be sought. FROM A DECLARATION OF PARTICIPANTS JOINTLY ADOPTED BY A COMMITTEE OF THE AMERICAN BAR ASSOCIATION AND A COMMITTEE OF PUBLISHERS.

LIBRARY OF CONGRESS CATALOGING-IN-PUBLICATION DATA

Nehter, Pedro.
 Theoretical analysis of high fuel utilizing solid oxide fuel cells / Pedro Nehter.
   p. cm.
 ISBN 978-1-60692-011-4 (softcover)
 1. Solid oxide fuel cells. I. Title. II. Title: Theoretical analysis of high fuel utilizing SOFCs.
 TK2931.N443 2008
 621.31'2429--dc22
                              2008033054

Published by Nova Science Publishers, Inc. New York

Contents

Preface		**vii**
Chapter 1	Introduction	1
Chapter 2	Analytical Solution of the Current Distribution	5
Chapter 3	Numerical Solution of the Current Distribution	19
Chapter 4	Nickel Oxide Formation at The Anode	33
Chapter 5	High Fuel Utilizing SOFC	37
Chapter 6	Conclusion	47
References		49
Index		51

PREFACE

The commercialization of fuel cells needs further developments in materials, power density and durability. These key issues are strongly related to the choice of electrochemical, thermodynamic and design parameters. This applies in particular to the sensitivity of the solid oxide fuel cell's (SOFC's) power density and durability. Achieving high power density has to be assessed carefully with regard to the cell's voltage, fuel utilization and efficiency.

The operation at high fuel utilization is particularly critical due to the decrement in the Nernst voltage and the formation of nickel oxide at nickel cermet anodes. Both effects are influenced by the local hydrogen to water ratio of the anode gas. Therefore, it is essential to understand the local resolution of the gas composition and its influence on the total power density. In this context, analytical solutions of the integral current density at a constant area specific resistance (ASR) are presented in this study.

As a result of the transferred species, loss mechanisms occur. These polarization losses are sensitively influenced by numerous mechanisms, which are strongly non-linear. Therefore, a finite difference model is chosen to analyze the influence of the major operational parameters on the power density. It is based on a two dimensional resolution of the local energy balance in the axial and radial direction of a tubular SOFC. This model includes heat transfer by conduction, convection and radiation as well as the heat sources due to the chemical and electrochemical reactions. The shift reaction and the reforming of residual methane are taken into account by a kinetic approach. The electrochemical losses of the hydrogen oxidation are determined by commonly used Butler-Volmer equation, binary diffusion, Knudsen diffusion and ohmic law.

Based on the finite difference simulation, a novel conceptual solution is proposed which allows improvements in the fuel utilization and power density.

The proposed configuration consists of an anode gas condenser which is used to increase the fuel utilization from 85% to 94-97% at a constant total cell area and higher efficiency. In this context, the tendency of the formation of nickel oxide at the anode is approximately estimated to take roughly the durability of the anode into account. This system could be applied to stationary CO_2 sequestering applications. Using methane as fuel, it is further shown that the CO_2 can be separated with a minimum demand of energy with the high fuel utilizing SOFC gas turbine cycle.

Chapter 1

INTRODUCTION

Previous studies have shown that solid oxide fuel cells (SOFC) promise highest system efficiencies [1,2]. This is mainly caused by the comparably low exergetic losses within the SOFC stack [3]. Siemens Westinghouse demonstrated the high durability of an atmospheric (system pressure ≈ 1.3 bar) SOFC system with a performance of 100 kW. The efficiency of such atmospheric SOFC systems mainly depends on the average cell voltage, the electrochemical fuel utilization and the demand of excess air. The fuel utilization is commonly chosen with 85% at Ni-Cermet anodes. A further amount in fuel utilization would result in a stronger formation of nickel oxide, which decreases the catalytic activity for the hydrogen oxidation. This effect is partly reversible but has to be avoided to increase availability and durability. To achieve appropriate system efficiencies, highest fuel utilizations are aspired in particular for SOFC systems which are not coupled with heat engines. In this context, it is necessary to consider the limitation mechanisms in fuel utilization with respect to the cell's power density and degradation mechanisms.

NOMENCLATURE

a	[$m^2\ s^{-1}$] / -	thermal conductivity coefficient / activity
A	[m^2]	Area
ASR	[Ωm^2]	area specific resistance
c_p	[$kJ\ mol^{-1}\ K^{-1}$]	heat capacity
C	[$W\ m^{-2}\ K^{-4}$]	radiation exchange factor
d	[m]	diameter
D	[$m^2\ s^{-1}$]	diffusion coefficient

NOMENCLATURE (CONTINUED)

E_N	[V]	Nernst voltage
F	[As mol^{-1}]	Faraday constant
G	[kJ mol^{-1}]	Gibbs free energy
GT		gas turbine
H	[kJ mol^{-1}]	enthalpy
i	[Am^{-2}]	electrical current density
$\bar{i}_{total}$	[Am^{-2}]	average electrical current density
I	[A]	electrical current
j_0	[A m^{-2}]	exchange current density
k	[mol/m^{2}bar^{2}s] / [J/K]	reaction rate coefficient / Boltzmann constant
K_p	-	equilibrium constant
M	[g mol^{-1}]	molar mass
Ni		nickel
$\dot{n}$	[mol s^{-1}]	molar flow
P	[W]	power
p_{el}	[Wm^{-2}]	area specific power
p	[bar]	partial pressure
$\dot{q}$	[W m^{-3}]	volumetric heat source
r	[m]	radius
R	[kJ mol^{-1} K^{-1}] / [Ω]	universal gas constant / ohmic resistance
$SOFC$		solid oxide fuel cell
S	[J mol^{-1} K^{-1}]	entropy
S/C	-	steam/carbon ratio
T	[K]	absolute temperature
u_f	-	fuel utilization
V	[V]	voltage
x	-	molar fraction
z	[m]	axial length

GREEK LETTERS

α	[W m^{-2} K^{-1}] / -	heat transfer coefficient / transfer coefficient
δ	[m]	layer thickness
η	-	efficiency
$\hat{\eta}$	[V]	polarization
ϑ	[°C]	temperature
$\bar{v}_A$	[m s^{-1}]	average velocity of the gas molecule
v_i	-	stoichiometric factor of component

GREEK LETTERS (CONTINUED)

$\Delta \nu$	-	stoichiometric difference of reaction
ρ	[kg m^{-3}]	density
τ	[s]	time
Ω_D	-	the collision integral

SUBSCRIPTS

Act	activation
Diff	diffusion
el	electrical
Elt	electrolyte
H$_2$	hydrogen
H$_2$O	water
i	component
irr	irreversible
Ca	cathode
Kn	Knudsen diffusion
loss	losses
O$_2$	oxygen
Pol	polarization
rad	radiation
reac	reaction
Ref	reforming reaction
rev	reversible
shift	shift reaction
Sys	system

Chapter 2

ANALYTICAL SOLUTION OF THE CURRENT DISTRIBUTION

2.1. POWER DENSITY

The total power P_{el} of a single cell with an equipotential cell area is given by the integral of the differential power.

$$P_{el} = V_{Cell} \cdot \int dI \tag{1}$$

The area specific power density p_{el} is a common parameter which is used to estimate the required cell area at a specific total power. It is determined by the quotient of the total power and the total cell area A, whereas the quotient of the total current I_{total} and the total cell area, which is calculated by the integral of differential currents dI, is equal to the average current density $\bar{i}_{total}$.

$$p_{el} = \frac{P_{el}}{A} = \frac{V_{Cell} \cdot \int dI}{A} = V_{Cell} \cdot \bar{i}_{total} \tag{2}$$

To solve this equation, it is necessary to consider the relation between the electrochemical conversion of the reacting species and the cell voltage. The electrochemical conversion of the reacting species in fuel cells is coupled directly with the exchanged electrical current. The molar consumption of reactants is determined by the Faradays law

$$\dot{n}_{H_2}^0 - \dot{n}_{H_2} = \frac{I_{total}}{2 \cdot F} \qquad \dot{n}_{H_2O} - \dot{n}_{H_2O}^0 = \frac{I_{total}}{2 \cdot F} \qquad \dot{n}_{O_2}^0 - \dot{n}_{O_2} = \frac{I_{total}}{4 \cdot F} \qquad (3)$$

where 2 mol electrons per mol hydrogen and 4 mol electrons per mol oxygen are exchanged. This implies that the electrolyte is free of electric leaks and the electrodes are free of any parallel reactions. The index "0" is used for the molar flow at the entry of the cell (Figure 1). The quotient of the converted hydrogen and the maximum convertible hydrogen is defined as fuel utilization uf.

$$uf = \frac{\Delta \dot{n}_{H2}}{\dot{n}_{H2}^0} = \frac{\dot{n}_{H2}^0 - \dot{n}_{H2}}{\dot{n}_{H2}^0} \qquad (4)$$

The maximum convertible hydrogen can also be expressed as maximum available electrical current I_{max},

$$I_{max} = F \cdot 2 \cdot \dot{n}_{H2}^0 \qquad (5)$$

$$I_{total} = uf \cdot I_{max} \qquad (6)$$

whereas the total electrical current I_{total} of the whole cell area A is proportional to the converted part of the maximum available current I_{max}. It is further assumed that the electrical current flow is exclusively directed perpendicular to the cell's area. Thus, ohmic losses in the direction parallel to the electrodes are neglected.

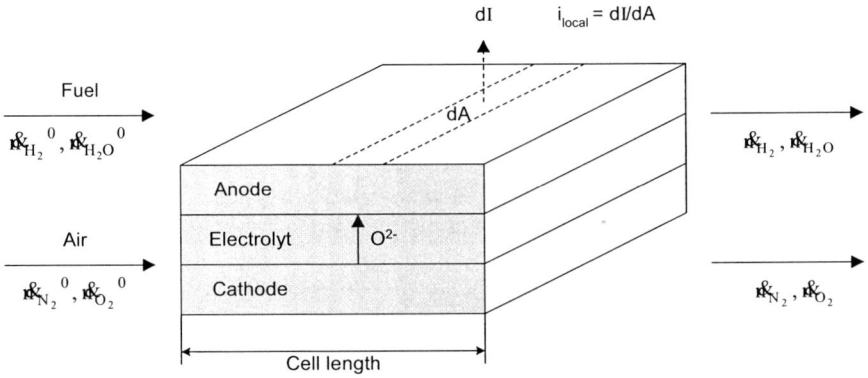

Figure 1. Schematic of a single solid oxide fuel cell.

In case of a non-linear distribution of the electrical current along the cell area, it is necessary to calculate the local current density i_{local},

$$i_{local} = \frac{dI}{dA} = \frac{I_{max} \cdot duf}{dA} \tag{7}$$

where the differential electrical current is equal to the differential utilization of the maximum current. In general, the existence of a current is caused by a potential gradient. In case of a fuel cell, the Nernst voltage E_N represents the driving force as potential gradient between the gaseous phases at the anode and the cathode. The Nernst voltage changes with the partial pressure of reacting species caused by the change in entropy. Thus, the influence of the fuel utilization on the Nernst voltage increases with higher operation temperatures.

$$E_{N,H2} = -\frac{1}{2 \cdot F}\left[\Delta^r G_{H2(T)} + T \cdot R \cdot \ln\left(\frac{p_{H2O}}{p_{H2} \cdot \sqrt{p_{O2}}} \cdot \sqrt{p_0}\right)\right] \tag{8}$$

As a result of the transferred species, loss mechanisms occur. These losses are well known as polarization losses in terms of the first law of thermodynamics. Polarization losses are sensitively influenced by numerous mechanisms which are strongly non-linear with respect to the operational parameters like the current density, electrical potentials, temperature, pressure, gas compositions and material properties. These parameters are assumed to be constant in case of a differential cell area, whereas the loss mechanisms are summarized in a constant area specific resistance ASR [Ωcm^2]. Thus, a change of the local overpotential ($E_{N(uf)}$ - V_{Cell}) at constant ASR complies with a proportional change in the local current density.

$$\frac{dI}{dA} = \frac{E_{N(uf)} - V_{Cell}}{ASR} \tag{9}$$

In general, high fuel utilizations are aspired to achieve high efficiencies, whereas different Nernst voltages occur along the cell area. Hence, the dependency of the fuel utilization from the electrical current (Eq. (7)) has to be implemented in Eq.(9).

$$\frac{I_{max} \cdot duf}{dA} = \frac{E_{N(uf)} - V_{Cell}}{ASR} \tag{10}$$

In this subsection, the ASR is assumed to be constant along the cell area to consider exclusively the influence of the Nernst voltage on the cell performance. Taking the local distribution of the current into account, this influence can be calculated by the integral of Eq. (10). This approach neglects the change in the Nernst potential caused by the diffusion in the gaseous bulk along the flow direction of reactants.

$$\int_{A=0}^{A=A_{total}} \frac{1}{ASR} dA = \int_{uf=0}^{uf=uf_{total}} \frac{I_{max}}{E_{N(uf)} - V_{Cell}} duf \tag{11}$$

Different conditions of SOFCs under test procedures and practical operation require different calculations for the evaluation of the test results. Thus, three cases at different distributions of the Nernst voltage are considered with regard to the solution of the integral cell area and integral fuel utilization, respectively.

Case A: Nernst voltage is constant
Case B: Nernst voltage changes inversely proportional with the fuel utilization
Case C: Nernst voltage changes according to Eq. (8)

The temperature of the gaseous phases and the SOFC are assumed to be equal and constant along the cell area.

<u>Case A:</u>
In case of comparably high maximum currents ($I_{max} \gg I$), the gaseous outlet composition is similar to the inlet composition. Thus, the Nernst voltage along the isothermal fuel cell is approximately constant. This condition occurs mostly at material related characterization tests, whereas small cell areas are investigated.

$$\begin{aligned} \dot{n}_{H2,out} &\approx \dot{n}_{H2}^0 \\ \dot{n}_{H2O,out} &\approx \dot{n}_{H2O}^0 \end{aligned} \quad \rightarrow \quad E_N \approx const. \tag{12}$$

The solution of Eq. (11) at a constant Nernst voltage shows a linear dependency of the cell voltage from the overpotential, represented by the fraction term of Eq. (13). The Ohmic law is similar to this fraction term, where the voltage drop is proportional to the current density represented by the quotient of the utilized part of the maximum current and the cell area.

$$V_{Cell} = E_N - \frac{ASR \cdot uf \cdot I_{max}}{A} \tag{13}$$

The cell voltage governed by Eq. (13) and Eq. (2) gives the power density at a constant

$$P_{el} = \left(E_N - \frac{ASR \cdot uf \cdot I_{max}}{A} \right) \cdot \frac{uf \cdot I_{max}}{A} \tag{14}$$

Case B:
If the fuel utilization occurs in a range of hydrogen to water pressure ratios (p_{H2}/p_{H2O}) between 0.7 and 0.3, the Nernst voltage changes approximately inversely proportional with the fuel utilization. Thus, a linear approach is used to determine the Nernst voltage from the fuel utilization,

$$E_{N(uf)} = E_N^0 + \frac{\Delta E_N}{\Delta uf} \cdot uf \tag{15}$$

where E_N^0 is the Nernst voltage at the entry of the anode and $\Delta E_N/\Delta uf$ is the slope of change in Nernst voltage.

$$\int_{A=0}^{A=A_{total}} \frac{1}{ASR} dA = \int_{uf=0}^{uf=uf_{total}} \frac{I_{max}}{E_N^0 + \frac{\Delta E_N}{\Delta uf} \cdot uf - V_{Cell}} duf \tag{16}$$

The solution for a linear Nernst voltage is given by

$$V_{Cell} = E_N^0 - \frac{\Delta E_N}{\Delta uf} \cdot uf \cdot \left[\exp\left(\frac{\frac{\Delta E_N}{\Delta uf} \cdot A}{ASR \cdot I_{max}} \right) - 1 \right]^{-1} \qquad (17)$$

and the power density is governed by

$$p_{el} = \left(E_N^0 - \frac{\Delta E_N}{\Delta uf} \cdot uf \cdot \left[\exp\left(\frac{\frac{\Delta E_N}{\Delta uf} \cdot A}{ASR \cdot I_{max}} \right) - 1 \right]^{-1} \right) \cdot \frac{uf \cdot I_{max}}{A} \qquad (18)$$

Case C:

In practical applications, high fuel utilizations result in low of hydrogen to water pressure ratios at the outlet of the anode, whereas high hydrogen pressures occur at the entry of the anode. Thus, the non-linear dependency of the Nernst voltage from the fuel utilization has to be taken into account with respect to the integral solution of Eq.(11).

$$\int_{A=0}^{A=A_{total}} \frac{1}{ASR} dA = \int_{uf=0}^{uf=uf_{total}} \frac{I_{max}}{-\frac{1}{2 \cdot F}\left[\Delta^r G_{H2(T)} + T \cdot R \cdot \ln\left(\frac{p_{H2O(uf)}}{p_{H2(uf)} \cdot \sqrt{p_{O2(uf)}}} \cdot \sqrt{p_0} \right) \right] - V_{Cell}} duf$$

(19)

The power density is determined by the numerical solution of Eq. (19) in combination with Eq.

A fixed cell area of 1 cm^2, a temperature of 800°C, a total pressure of 1 bar, an oxygen partial pressure of 0.21 bar and an area specific resistance of 1 Ωcm^2 are chosen to compare the cell performances at uniform conditions. The total current I_{total} = 0.3A and the molar hydrogen fraction at the outlet of the anode x_{H2} = 0.299 are kept constant as well. Hence, the hydrogen flow rate and the hydrogen fraction at the entry of the anode are adjusted to obtain the linear and non-linear dependency of Nernst voltage from the fuel utilization. The calculation results of each case are summarized in Table 1.

Table 1. Analytical calculation results of the power density

Case	$\dot{n}_{H2}^{0}$ [10^{-6} mol/s]	x_{H2}^{0} [mol/mol]	I_{max} [A]	uf [%]	V_{Cell} [V]	p_{el} [W/cm^2]
A: E_N = const	500	0.3	96.4	0.31	0.698	0.209
B: $E_N \sim (x_{H2}/x_{H2O})$	2.72	0.7	0.524	57.1	0.726	0.217
C: $E_N = f(x_{H2}/x_{H2O})$	2.22	0.99	0.428	70	0.746	0.224

Case A is obtained at a high hydrogen flow rate, which results in a high maximum current, low fuel utilization and approximately constant Nernst Voltage. The local resolution of the Nernst voltage and the molar fraction of the hydrogen (Figure 2) shows a constant local distribution. The locally constant overpotential of 0.29V results in a cell voltage of 0.698V. Thus, the current and power density obtain locally constant values with i_{local} = 0.3 Acm^{-2} and $p_{el,local}$ = 0.209 Wcm^{-2} in each section of the cell's area (Figure 3).

In Case B, the hydrogen flow rate is chosen with 2.72·10^{-6} mol/s to obtain a change in the molar hydrogen fraction from 0.7 at the entry to 0.299 at the outlet of the anode. In this range of molar fraction, the Nernst voltage changes approximately inversely proportional with the fuel utilization. Even if the Nernst voltage, the overpotential, the current density and the power density change inversely proportional with the fuel utilization, the local distribution of the Nernst voltage and the current density show a slightly non-linear dependency. This is caused by the fact that the area section, which is required to transfer a specific current at a specific overpotential through the cell, increases disproportionately with lower overpotentials (Eq. (9). It is further shown that the differential current is displaced to the entry of the cell, where higher overpotentials occur. To keep the total current constant, the differential current at the outlet of the cell has to be lower than the total or average current density. Even if the Nernst voltage at the outlet of the cell is kept constant, the cell voltage of case B achieves a value of 0.726V, which is 28mV higher than the cell voltage of case A. This is caused by a higher average Nernst voltage in case B compared to case A. Thus, the average power density increases as well.

In Case C, the hydrogen flow rate is chosen with 2.22·10^{-6} mol/s to obtain a non-linear change of the Nernst voltage at a variation of the molar hydrogen fraction from 0.99 at the entry to 0.299 at the outlet of the anode. The Nernst voltage, the overpotential, the current density and the power density change disproportionately with the fuel utilization in particular at the entry of the anode. This is mainly caused by the change in entropy, which achieves particularly high absolute values at $p_{H2O}/p_{H2} \gg 1$ or $p_{H2O}/p_{H2} \ll 1$. One of these conditions occurs at the entry of the anode of the case C. Thus, the differential current is displaced disproportionately to the entry of the cell. This results in an amount of the cell voltage up to 0.746V, which is 48mV higher compared to the cell voltage of case A.

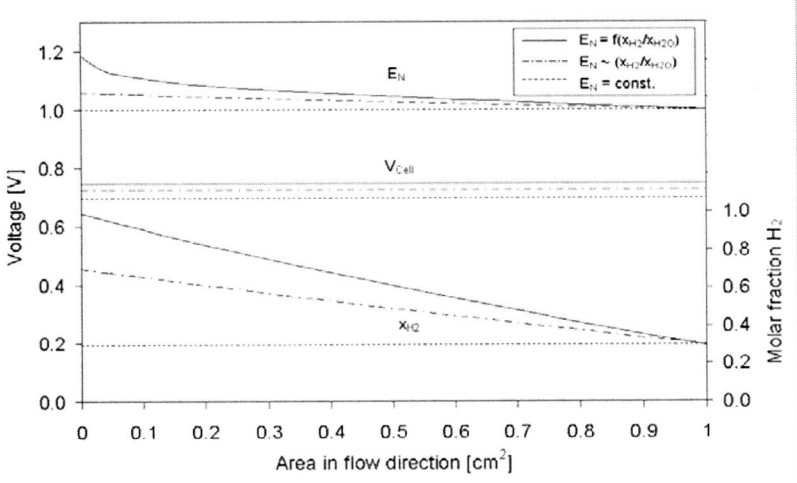

Figure 2. Local resolution of the Nernst voltage and the molar hydrogen fraction.

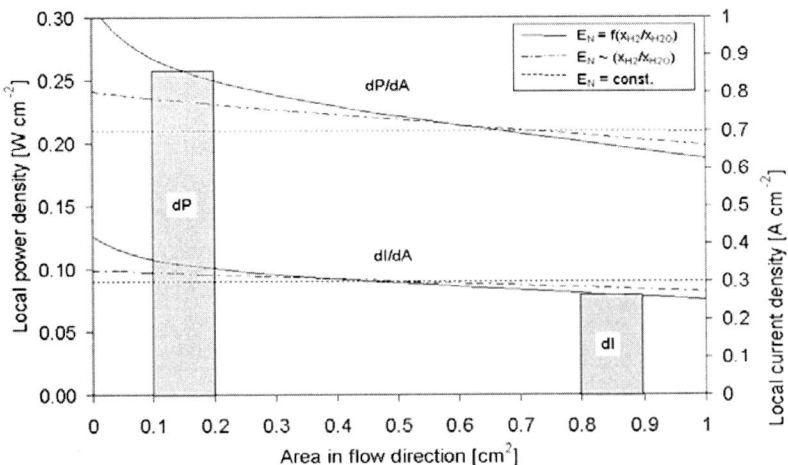

Figure 3. Local resolution of the power and the current.

The results are not unexpected from a qualitative point of view. As long as the Nernst voltage is considered to be the driving force, the cell performance will increase with a higher average Nernst voltage or higher reversible power P_{rev} at a constant ASR and constant total current density. The reversible power given by the Gibbs free energy of reaction can be calculated independently from the local current distribution as follows:

$$P_{rev} = I_{max} \int_{uf=0}^{uf=uf_{total}} E_{N(uf)}\, duf = uf \cdot I_{max} \cdot \overline{E}_{N(uf)} \qquad (20)$$

The quotient of the total power and reversible power at operating temperature gives the relation between the converted power at irreversible conditions and the maximum convertible power of the fuel cell at reversible conditions.

$$\eta_B = \frac{P_{el}}{P_{rev}} = \frac{P_{rev} - P_{loss}}{P_{rev}} = 1 - \frac{P_{loss}}{P_{rev}} \qquad (21)$$

The comparison of the cases A, B and C (Table 2) shows that the total power P_{el} increases non-linear and disproportionate with a higher reversible power. Thus, the efficiency η_B changes disproportionately as well. On the one hand, is this caused by the change of the average Nernst voltage and on the other hand by different local distributions of the current.

Table 2. Analytical calculation results of the losses

Case	P_{rev} [W]	$\overline{E}_{N(uf)}$ [V]	V_{Cell} [V]	P_{el} [W]	P_{loss} [W]	η_B [%]
A: E_N = const	0.299	0.998	0.698	0.209	0.0899	69.9
B: $E_N \sim (x_{H2}/x_{H2O})$	0.307	1.026	0.726	0.217	0.0902	70.6
C: $E_N = f(x_{H2}/x_{H2O})$	0.315	1.050	0.746	0.224	0.0912	71.0

Some models use the average Nernst voltage, according to Eq. (13), to determine the cell voltage and cell power from a given ASR and current density. The results of this approach are in deviation to the results of the integral determination according to Eq.(11).

$$P_{el,case B, case C} \neq uf \cdot I_{max} \cdot \left(\overline{E}_N - ASR \cdot \overline{i}_{total} \right) \qquad (22)$$

The relative error of the total power calculated by the average Nernst voltage related to the integral results is err$_{case A}$ = 0%, err$_{case B}$ = 0.1% and err$_{case C}$ = 0.5%. Thus, the average Nernst voltage can be used to determine the cell power as long as the conditions of case A are complied. At conditions according to case B and case C the integral determination of the cell's performance (Eq.(11)) is recommended.

2.2. REQUIRED CELL AREA

The total required cell area for an isothermal cell with a constant ASR is calculated as follows:

$$A = I_{max} \cdot ASR \cdot \underbrace{\int_{uf1}^{uf2} \frac{1}{-\frac{1}{2 \cdot F}\left[\Delta^r G_{H2(T)} + T \cdot R \cdot \ln\left(\frac{p_{H2O(uf)}}{p_{H2(uf)} \cdot \sqrt{p_{O2(uf)}}} \cdot \sqrt{p_0}\right)\right] - V_{Cell}} duf}_{= uf/\overline{\eta} = uf/(ASR \cdot \overline{i}_{total})} \qquad (23)$$

The integral part of Eq. (11) represents the inverse average overpotential which is proportional to the inverse total current density at constant ASR.

$$\frac{uf}{\overline{\eta}} = \frac{uf}{ASR \cdot \overline{i}_{total}} = \int_{uf1}^{uf2} \frac{1}{-\frac{1}{2 \cdot F}\left[\Delta^r G_{H2(T)} + T \cdot R \cdot \ln\left(\frac{p_{H2O(uf)}}{p_{H2(uf)} \cdot \sqrt{p_{O2(uf)}}} \cdot \sqrt{p_0}\right)\right] - V_{Cell}} duf \qquad (24)$$

Furthermore, an average Nernst potential can be defined taking the integral overpotential into account, which is inversely proportional to the total required cell area.

$$\overline{\hat{\eta}} + V_{Cell} = \overline{E}_N \tag{25}$$

If the electrical performance of the cell P_{el}, the average cell voltage V_{Cell} and the fuel utilization uf are requested parameters,

$$P_{el} = I_{total} \cdot V_{Cell} = uf \cdot I_{max} \cdot V_{Cell} \tag{26}$$

the required cell area can be calculated by Eq. (11) and Eq.(

$$A = \frac{P_{el}}{uf \cdot V_{Cell}} \cdot ASR \cdot \int_{uf1}^{uf2} \frac{1}{-\frac{1}{2 \cdot F} \left[\Delta^r G_{H2(T)} + T \cdot R \cdot \ln\left(\frac{p_{H2O(uf)}}{p_{H2(uf)} \cdot \sqrt{p_{O2(uf)}}} \cdot \sqrt{p_0} \right) \right] - V_{Cell}} duf \tag{27}$$

The required cell area and the efficiency are calculated for an exemplary power of 1W (Figure 4). The efficiency of the cell is defined as a quotient of the cell's power and the enthalpy of reaction. Figure 4 shows that the required cell area increases with higher cell voltages and fuel utilizations. As a result, the fuel consumption decreases due to the higher electrical fuel cell efficiency. This is an opposite effect with regard to the investment and fuel cost. Furthermore, the long term degradation of the cell's power decreases with higher cell voltages and lower fuel utilzations. Thus, the choice of operational parameters at the design point has to be assessed carefully for each application and for the particular state of the art.

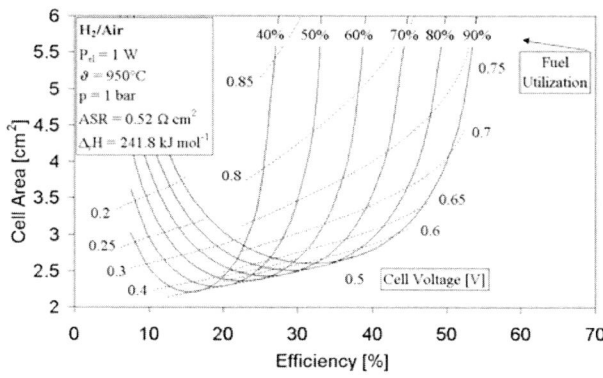

Figure 4. Required cell area.

Figure 5 shows the range of local current densities which vary due to the change in the Nernst voltage. Here, a typical V-i-characteristic (p = 1 bar, p_{O2} = 0.21 bar, T = 1223 K, Fuel: Hydrogen, uf = 85%) is chosen as a reference (black graph, Figure 5). The cell's temperature and the local ASR are assumed to be constant.

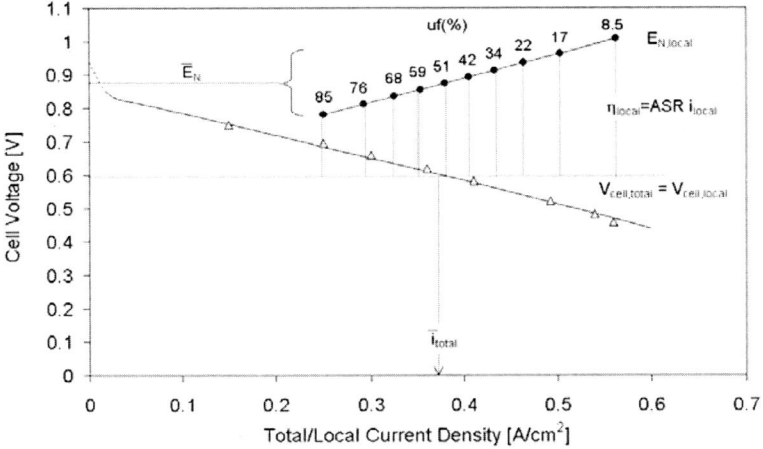

Figure 5. Total and local current density at a cell voltage of 0.6V

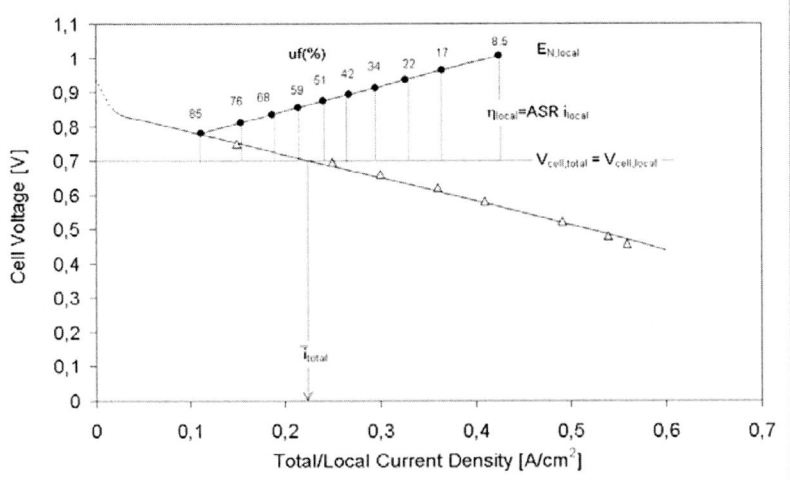

Figure 6. Total and local current density at a cell voltage of 0.7V.

The average Nernst voltage calculated by Eq.(24) and Eq.(25) is 0.87V at a fuel utilization of 85% and a cell voltage of 0.6V (Figure 5). This implies an average overpotential of 0.27V, whereas the ASR is calculated by Eq.(27) to 0.722Ωcm^2 at the specified total current density of 0.375 Acm^{-2}. The reverse calculation results at a different cell voltage of 0.7V is shown in Figure 6.

The local current densities are displaced to lower values, which is caused by the decreasing overpotential in average due to the higher cell voltage. Here, the average Nernst voltage is 0.86V and the average overpotential 0.16 V. The total current density of 0.22 Acm^{-2} calculated by the total required cell area and the total current shows a good congruence with the V-i-characteristics at a cell voltage of 0.7 V. This implies that the V-i-characteristics, measured at constant fuel utilizations, can be used to estimate the required cell area at different cell voltages at the design point.

Chapter 3

NUMERICAL SOLUTION OF THE CURRENT DISTRIBUTION

3.1. SOFC CONCEPT

Tubular SOFC concepts give the opportunity to implement heat sinks like preheater and reformer inside the cell and the stack. Thus, the excess air demand, which is required for the cooling of the cell, can be reduced. Thermodynamic and economic analysis have shown that low excess air demands and stack-integrated heat sinks are prerequisites to achieve low cost and efficient SOFC systems [3]. Thus, an air preheater tube is considered for the SOFC concept of this study (Figure 7).

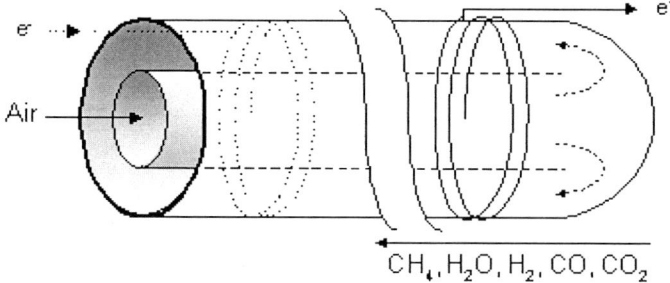

Figure 7. Tubular SOFC.

In this study, a two dimensional resolution of the local energy balance in the axial and radial direction of a tubular SOFC is applied. The model uses the

Cranck-Nicolson discretization, whereas the generated 2D-tridiagonal matrix is solved iteratively by the Gauss-Seidel method. The model includes the calculation of chemical reactions like the steam reforming, shift reaction, electrochemical reactions and heat transfer. The thermal part of the model calculates the electrical output as well as the heat sources from the electrochemical part of the model and determines a new cell temperature. The overpotential and the Kirchhoff's law are used to determine the local current density and local cell voltage.

3.2. ENERGY BALANCE

The loss mechanisms of a SOFC are strongly temperature sensitive. Thus, the local temperature distribution has to be calculated to determination of the cell's power. Heat transfer (Figure 8) occurs through thermal conductivity, convection (q_α) and radiation (q_{rad}). The conductive heat transfer in the solid material of the cell can be calculated by solving the local energy balance.

$$\frac{\partial T}{\partial \tau} = a \cdot div\ grad\ T + \frac{\dot{q}}{\rho \cdot c_p} \tag{28}$$

Here T is the temperature, τ the time, ρ the density, c_p the specific heat, a the thermal conductivity coefficient of the solid ceramic of the SOFC and q the volume specific heat source. It is assumed that the thermal conductivity coefficient is constant in each layer and each direction.

$$div\ grad\ T = \frac{1}{r}\frac{\partial T}{\partial r} + \frac{\partial^2 T}{\partial r^2} + \frac{1}{r^2}\frac{\partial^2 T}{\partial \varphi^2} + \frac{\partial^2 T}{\partial z^2} \tag{29}$$

The differential equation is applied in axial and radial direction of the cell. The temperature profile in the circumferential direction is assumed to be distributed homogenous.

The heat sources can be divided into the reaction enthalpy of the reforming, the shift reaction ($q_{ref+shift}$) and the reversible heat ($q_{ox,rev}$) of fuel oxidation.

$$\dot{q}_{reac} = \frac{1}{\Delta r \cdot \Delta A} \cdot \left(\Delta \dot{H}_{ref} + \Delta \dot{H}_{shift} + \Delta \dot{Q}_{ox,rev}\right) \tag{30}$$

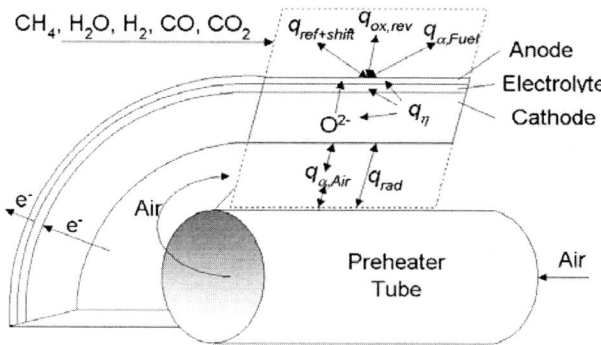

Figure 8. Schematics of heat transfer.

This reaction heat is equal to the difference of the all over reaction enthalpy and the reversible work of fuel oxidation.

$$\dot{q}_{reac} = \frac{1}{\Delta r \cdot \Delta A} \cdot \left(\Delta \dot{H}_{read(T=const)} - \dot{W}_{ox,rev} \right) \qquad (31)$$

The oxidation of hydrogen and carbon monoxide are assumed to be the exclusive electrochemical reactions. It can be shown that the reversible work of these two reactions have the same value if the shift reaction complies its equilibrium condition. The different mechanism losses of the hydrogen and carbon monoxide oxidation are neglected to determine the reversible work from one of these reactions. Convective heat transfer occurs between solid surfaces and the gas streams

$$\dot{q}_\alpha = \frac{\alpha(Nu)}{\Delta r} \cdot \left(\vartheta_{solid} - \vartheta_{fluid} \right) = \frac{1}{\Delta r \cdot \Delta A} \cdot \Delta \dot{H}_{fluid} \qquad (32)$$

where Δr the radial segment and ΔA the area segment perpendicular to flow direction of the fuel and the air.

It is assumed that radiation transfer occurs exclusively between the cathode and the air preheater tube.

$$\dot{q}_{rad} = \frac{C_{Ca,ft}}{\Delta r} \cdot \left[\left(\frac{T_{Ca}}{100} \right)^4 - \left(\frac{T_{ft}}{100} \right)^4 \right] \qquad (33)$$

C is the radiation exchange factor between the cathode and the preheater tube.

The heat sources due to the polarization effects (q_η) are determined by

$$\dot{q}_\eta = \frac{\hat{\eta} \cdot i \cdot \Delta A_i}{\Delta r \cdot \Delta A} = \frac{R \cdot (i \cdot \Delta A_i)^2}{\Delta r \cdot \Delta A} = \frac{\delta \cdot R_\delta \cdot i^2 \cdot \Delta A_i}{\Delta r \cdot \Delta A} \tag{34}$$

where $\hat{\eta}$ is the overpotential, i the current density, ΔA_i the area segment in the equivalent electrical current flow direction, R the equivalent ohmic resistance in Ω, R_δ the equivalent specific ohmic resistance in Ωm and δ the layer thickness in flow direction of the current.

3.3. MASS BALANCE

For the calculation the fluid channel is divided into slices of discrete distances. The gas compositions are assumed to be homogenous distributed in each slice of the ideal fluids. The considered chemical reactions are:

$CH_4 + H_2O \leftrightarrow 3H_2 + CO$ Steam Reforming

$CO + H_2O \leftrightarrow CO_2 + H_2$ Shift Reaction

$H_2 + 0.5\, O_2 \leftrightarrow H_2O$ Hydrogen Oxidation

The shift is assumed to comply its equilibrium composition.

$$\frac{x_{H_2} \cdot x_{CO_2}}{x_{CO} \cdot x_{H_2O}} = \exp\left(-\frac{\Delta_r H - T \cdot \Delta_r S}{R \cdot T}\right) \tag{35}$$

where x is the molar fraction of reacting species, $\Delta_r H$ is the reaction enthalpy and $\Delta_r S$ is the reaction entropy.

The reforming reaction is kinetically controlled. Some investigations [4] at Ni-Cermet anodes have shown that the surface boundary diffusion and pore diffusion can be neglected below 700°C and the surface boundary diffusion can be neglected between 800°C and 900°C [5]. In this study both transport processes were neglected. The conversion of methane is given [6] by

$$\dot{n}_{CH_4,0} - \dot{n}_{CH_4} = \Delta A_{Kat} \cdot k_{eff} \cdot p_{CH_4} \cdot p_{H_2O} \tag{36}$$

where ΔA_{Kat} is the catalytic area calculated by the geometry of the anode, k_{eff} the effective rate coefficient and p the partial pressure of gas species. The conversion rate coefficient is determined by the Arrhenius equation

$$k_{eff} = k_0 \cdot \exp\left(-\frac{E_{Act}}{R \cdot T}\right) \tag{37}$$

where k_0 is the pre-exponential coefficient and E_{Act} the activation energy. The limiting conversion is given by the thermodynamic equilibrium:

$$\frac{x_{H_2}^3 \cdot x_{CO}}{x_{CH_4} \cdot x_{H_2O}} = \left(\frac{p^0}{p}\right)^2 \cdot \exp\left(-\frac{\Delta_r H - T \cdot \Delta_r S}{R \cdot T}\right) \tag{38}$$

The local conversion of the hydrogen oxidation is determined by Faradays law as follows:

$$\dot{n}_{H_2,0} - \dot{n}_{H_2} = \frac{i \cdot \Delta A}{2 \cdot F} \tag{39}$$

$$\dot{n}_{O_2,0} - \dot{n}_{O_2} = \frac{i \cdot \Delta A}{4 \cdot F} \tag{40}$$

3.4. ELECTROCHEMICAL APPROACHES

Polarization is a well known parameter in the analysis of electrochemical devices. To calculate the relationship between local voltage $V_{Cell,L}$ and local current density the following general expression can be used:

$$V_{Cell,L} = E_{N,L} - \hat{\eta}_{Act,An} - \hat{\eta}_{Act,Ca} \\ - \hat{\eta}_{Diff,An} - \hat{\eta}_{Diff,Ca} - \hat{\eta}_{\Omega,Elt} \tag{41}$$

The overpotential can be divided in activation, diffusion (concentration) and ohmic overpotential. The Nernst voltage is considered to be the driving force, which is a function of the reversible work of the hydrogen oxidation (Eq.(8)).

In this study, the ohmic voltage drop is calculated iteratively taking the Kirchhoff's law into account.

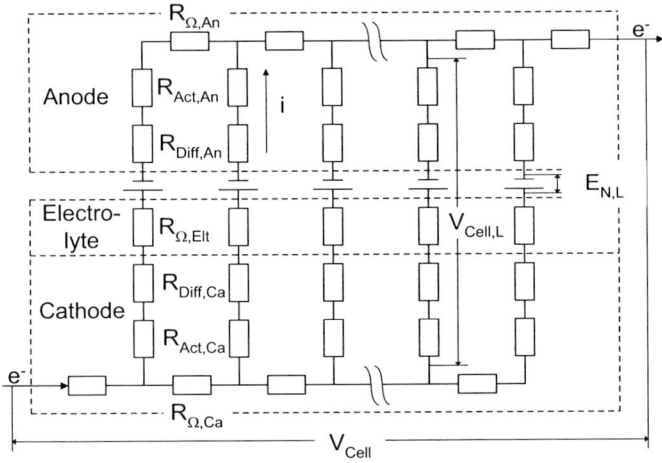

Figure 9. Equivalent electrical circuit.

3.5. ACTIVATION OVERPOTENTIAL

The activation of a chemical reaction involves energy barriers which must be overcome by the reacting species. These energy barriers are called the activation energy and result in charge transfer polarization. The activation polarization of an electrochemical reaction is commonly expressed by the Butler-Volmer equation

$$i = i_0 \left[\exp\left(\frac{\hat{\alpha} \cdot z \cdot F \cdot \hat{\eta}_{Act}}{R \cdot T} \right) - \exp\left(-\frac{(1-\hat{\alpha}) \cdot z \cdot F \cdot \hat{\eta}_{Act}}{R \cdot T} \right) \right] \quad (42)$$

where $\hat{\alpha}$ is the transfer coefficient and i_0 the exchange current density. The transfer coefficient is approximately 0.5 for fuel cells. The exchange current density is the cathodic and anodic electrode reaction rate at the equilibrium potential. A high exchange current density results in a good performance. With $\hat{\alpha} = 0.5$ the activation polarization can be simplified as follows

$$\hat{\eta}_{Act} = \frac{R \cdot T}{0{,}5 \cdot z \cdot F} \cdot \operatorname{arsinh}\left(\frac{i}{2 \cdot i_0}\right) \qquad (43)$$

For the anodic exchange current density of the hydrogen oxidation at Ni-YSZ anodes, Mogensen [7] proposed

$$i_{0,An} = \gamma_{An} \cdot \left(\frac{p_{H_2}}{p_{0,An}}\right) \cdot \left(\frac{p_{H_2O}}{p_{0,An}}\right) \cdot \exp\left(-\frac{E_{Act,An}}{R \cdot T}\right) \qquad (44)$$

where $E_{Act,An}$ is the activation energy, γ_{An} an empirical pre-exponential factor, p_{H2} the partial pressure of hydrogen, p_{H2O} the partial pressure of water and $p_{0,An}$ the reference pressure at the anode.

The cathodic exchange current density of LSM cathodes is expressed as [8]:

$$i_{0,Ca} = \gamma_{Ca} \cdot \left(\frac{p_{O_2}}{p_{0,Ka}}\right)^{0{,}25} \cdot \exp\left(-\frac{E_{Act,Ca}}{R \cdot T}\right) \qquad (45)$$

3.6. Concentration Overpotential

The concentration overpotential at electrodes is a fictive electrical potential drop caused by the concentration difference in the gaseous phase. Diffusion through porous materials is typically described by the binary molecular diffusion and the Knudsen diffusion. In this study, both types of diffusion are taken into account to estimate the concentration overpotential. It is assumed that the concentration gradient along the pore is much higher than the concentration gradient within the surface boundary. Thus, the concentration gradient inside the surface boundary is neglected.

For the ideal gas model the anodic concentration overpotential which is proportional to the irreversible entropy production can be written as

$$\hat{\eta}_{Diff,An} = \frac{R \cdot T}{z \cdot F} \cdot \ln\left(\frac{p_{H_2}^*}{p_{H_2}^0} \cdot \frac{p_{H_2O}^0}{p_{H_2O}^*}\right) = \frac{T}{z \cdot F} \cdot \Delta S_{irr,An} \qquad (46)$$

and for the cathodic concentration overpotential as follows

$$\hat{\eta}_{Diff,Ca} = \frac{R \cdot T}{2 \cdot z \cdot F} \cdot \ln\left(\frac{p^*_{O_2}}{p^0_{O_2}}\right) = \frac{T}{2 \cdot z \cdot F} \cdot \Delta S_{irr,Ca} \qquad (47)$$

where p^0 is the partial pressure in the gas bulk, p^* the partial pressure at the end of the pores. The concentration profiles inside the pores can be estimated by using the Fick's law (Eq.(48) - Eq.(50)). The molar fraction x can be expressed as

$$x^*_{H_2} = x^0_{H_2} - \frac{R \cdot T \cdot i \cdot \delta_{An}}{p \cdot 2 \cdot F \cdot D_{eff,An}} \qquad (48)$$

$$x^*_{H_2O} = x^0_{H_2O} + \frac{R \cdot T \cdot i \cdot \delta_{An}}{p \cdot 2 \cdot F \cdot D_{eff,An}} \qquad (49)$$

for the equimolar diffusion of hydrogen and water and

$$x^*_{O_2} = 1 + \left(x^0_{O_2} - 1\right) \cdot \exp\left(\frac{R \cdot T \cdot i \cdot \delta_{Ca}}{p \cdot 4 \cdot F \cdot D_{eff,Ca}}\right) \qquad (50)$$

for the self-diffusion of oxygen in nitrogen. δ_{An} is here the thickness of the anode, δ_{Ca} the thickness of the cathode and D_{eff} the effective diffusion coefficient. In this model, the effective diffusion coefficient takes the Knudsen diffusion as well as the ordinary diffusion at the cathode and the binary diffusion at the anode into account.

The binary diffusion coefficient between gas species A and B in free space is given by the Chapman–Enskog theory

$$D_{AB} = 1{,}8583 \cdot 10^{-6} \cdot \left(\frac{1}{M_A} + \frac{1}{M_B}\right)^{1/2} \cdot \frac{T^{3/2}}{p \cdot \sigma^2_{AB} \cdot \Omega_D} \qquad (51)$$

where σ is the collision diameter (Lennard–Jones length) and Ω_D is the collision integral based on the Lennard–Jones potential.

The Knudsen diffusion coefficient for a species A is defined as

$$D_{Kn,A} = \frac{2 \cdot R_{Por} \cdot \bar{v}_A}{3} \tag{52}$$

where R_{Por} is the pore radius and $\bar{v}$ is the average velocity of the gas molecule. Both ordinary diffusion and Knudsen diffusion can be combined within the effective diffusion coefficient which can be written as:

$$\frac{1}{D_{eff,Por,A}} = \frac{\tau_p}{\theta} \cdot \left(\frac{1}{D_{AB}} + \frac{1}{D_{Kn,A}} \right) \tag{53}$$

The tortuosity τ_P and porosity θ are used to account the geometry of the pores and relation between solid space and pore space of the electrode.

If the presence of other species except hydrogen and water inside the anodic pores is neglected, the anodic effective diffusion coefficient can be expressed for the ideal gas-mixture as [9]:

$$D_{eff,An} = \left(\frac{p_{H_2O}}{p_{ges}} \right) \cdot D_{eff,Por,H_2} + \left(\frac{p_{H_2}}{p_{ges}} \right) \cdot D_{eff,Por,H_2O} \tag{54}$$

Inside the cathodic pores the cathodic effective diffusion coefficient can be formulated as:

$$D_{eff,Ca} = D_{eff,Por,O_2} \tag{55}$$

3.7. OHMIC OVERPOTENTIAL

Ohmic losses occur according to the flow of ions in the electrolyte

$$\hat{\eta}_{\Omega,Elt} = i \cdot R_{\Omega,Elt} \tag{56}$$

and the flow of electrons through the electrode materials.

$$\hat{\eta}_{\Omega,An} = i \cdot R_{\Omega,An} \tag{57}$$

$$\hat{\eta}_{\Omega,Ca} = i \cdot R_{\Omega,Ca} \tag{58}$$

The ionic conductivity of the electrolyte depends strongly on material, temperature T and thickness δ. For YSZ electrolytes Achenbach [10] has proposed the following dependency:

$$R_{\Omega,Elt} = \delta_{Elt} \cdot \frac{1\,\Omega m}{33400 \cdot \exp\left(-\dfrac{10300K}{T}\right)} \tag{59}$$

In this study the electronic resistances of the anode and cathode are assumed to be constant.

3.8. Solution Technique

Figure 10 shows a flow chart of the calculation method. The model includes the calculation of chemical reactions like the steam reforming, shift reaction, electrochemical reactions and thermal heat transfer.

The overpotential and the Kirchhoff's law are used to determine the local current density and local cell voltage within the electrochemical part of the model. The local cell currents are summed up to the total cell current. For the calculation of steady state conditions the calculated cell voltage is compared with the specified cell voltage. The cell's heat sources are calculated from the reversible change of entropy due to the oxidation of hydrogen, the local overpotential, the heat transfer by radiation and the heat transfer by convection which includes the sensible enthalpy changes in the gaseous phases. The heat sources are referred to the solid nodes of the thermal SOFC model. The transient model uses the Cranck-Nicolson discretization (axial:51 nodes, radial: 7 nodes) which has the advantage that no stability boundaries have to be considered. The generated 2D-tridiagonal matrix is solved iteratively by the Gauss-Seidel method.

3.9. Exemplary Results

In this subsection, the current and temperature distribution of the tubular SOFC is calculated exemplary. The initial parameters of this calculation are listed

Table 3. The temperatures of the fuel and the oxidant at the entry of the anode and cathode are assumed to be equal.

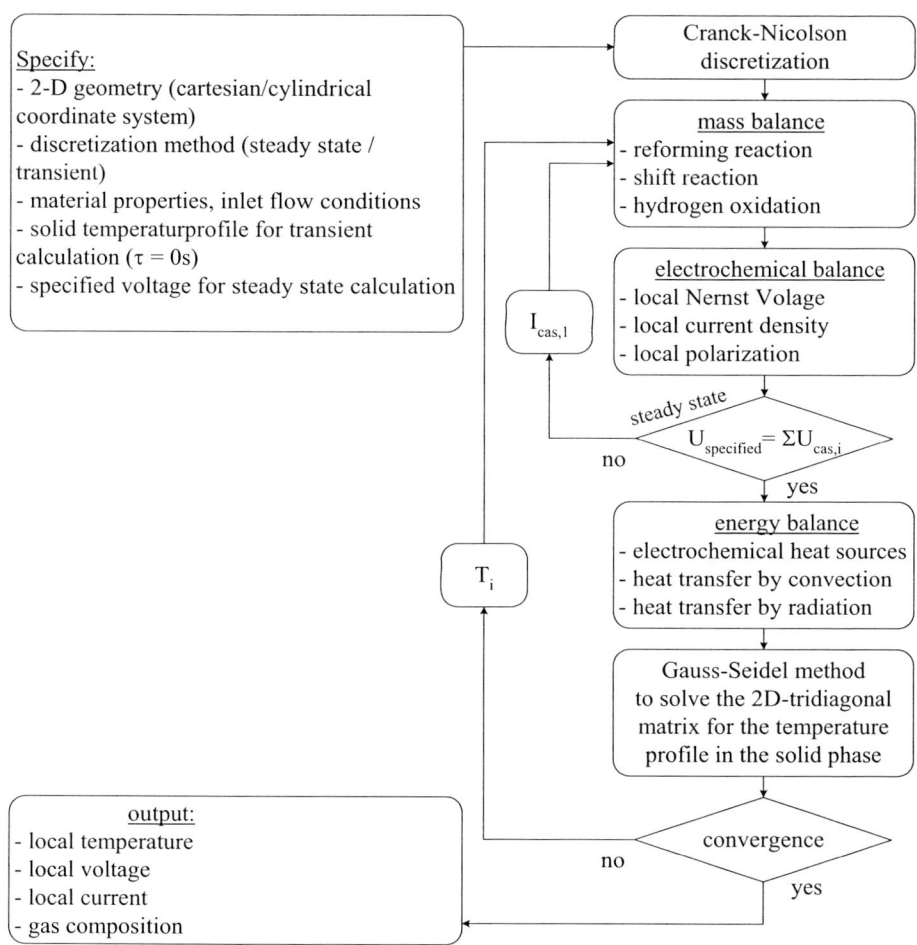

Figure 10. Flowchart of solution technique.

Table 3. Initial parameters

Cell length [m]	0.1
Cell inside diameter [m]	$2 \cdot 10^{-3}$
Thickness anode (Ni-YSZ) [m]	$0.05 \cdot 10^{-3}$
Thickness cathode($La_{0.6}Sr_{0.4}MnO_3$)[m]	$0.05 \cdot 10^{-3}$
Thickness electrolyte (YSZ) [m]	$0.1 \cdot 10^{-3}$
Heat capacity ceramic [J $kg^{-1} K^{-1}$]	400
Heat conduction coefficient ceramic [W $m^{-1} K^{-1}$]	2
Density [kg m^{-3}] ceramic	6600
Air preheater tube diameter [m]	$1 \cdot 10^{-3}$
Thickness Air preheater tube [m]	$0.1 \cdot 10^{-3}$
Heat conduction coefficient air preheater tube [W $m^{-1} K^{-1}$]	6
pre-exponential coefficient i_0 anode[A m^{-2}]	$18.9 \cdot 10^9$
pre-exp. coefficient i_0 cathode [A m^{-2}]	$1.034 \cdot 10^9$
Activation energy anode [kJ mol^{-1}]	120
Pore radius; Tortuosity; Porosity anode [m; - ; %]	$0.5 \cdot 10^{-6}$; 6; 30
Activation energy cathode [kJ mol^{-1}]	130
Pore radius; Tortuosity; Porosity cathode [m; - ; %]	$1.2 \cdot 10^{-6}$; 6; 50
Activation energy ref. [kJ mol^{-1}]	150
Ohmic restistance anode/cathode [Ωm]	$5.6 \cdot 10^{-6}$
Radiation exchange factor [W $m^{-2} K^{-4}$]	0.1059
Temperature gas inlet [K]	973.15
Pressure gas inlet [bar]	1.04
Fuel inlet flow [mol s^{-1}]	$2.457 \cdot 10^{-6}$
Fuel molar fraction (equal to a pre-reforming rate of 30%)	0.17 CH_4, 0.48 H_2O, 0.26 H_2, 0.02 CO, 0.04 CO_2
Fuel Steam/Carbon ratio	2.85
Fuel/Air flow time constant [s]	2/-
Air inlet flow (dry air) [mol s^{-1}]	$8.282 \cdot 10^{-5}$
Average cell voltage [V]	0.7
Fuel utilization [%]	85%

The molar fraction and temperature distribution of the solid ceramic along the flow direction (z-direction from fuel inlet) is shown in Figure 11. In the first section of the fuel cell the hydrogen fraction increases due to the endothermic reforming of methane. This has a cooling effect at the fuel's entry. The hydrogen fraction is decreased by the electrochemical oxidation of the hydrogen which is proportional to the local current density (Figure 12). Thereupon, the shift reaction produces hydrogen as long as carbon monoxide is available.

The local current density increases along the flow direction as a consequence of the low temperatures and high polarization at the fuel inlet of the cell. Even if the temperature increases up to a cell length of 7 cm, the local current density decreases from a cell length of 4 cm. This is caused by the losses due to the Nernst voltage and concentration gradients, which are in sum higher than the change in the temperature-sensitive losses.

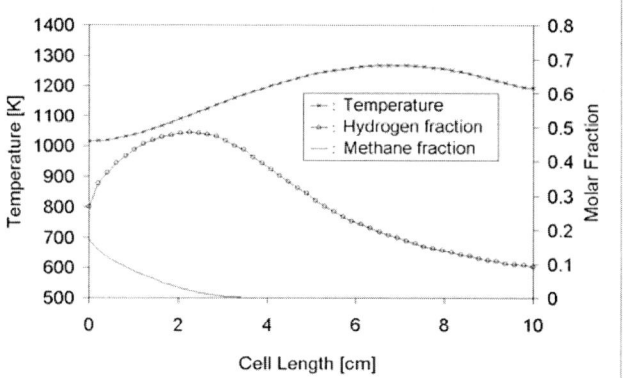

Figure 11. Temperature of the tubular SOFC.

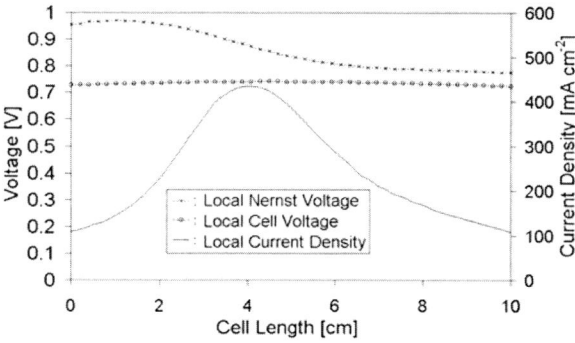

Figure 12. Local voltage and current density of the tubular SOFC.

The maximum temperature of 1268 K is located at a cell length of 7cm. The temperature drop in the last third of the cell length is caused by the strong amount in radiative and convective heat transfer between the air preheater tube and the cathode. The local cell voltage shows a slight maximum in the middle of the cell length which can be explained by ohmic losses in the anode and cathode. This is because the current collectors (Figure 9) lead the highest current density at the beginning of the anode (cell length: 0cm) and at the end of cathode (cell length:10 cm). Under these conditions the single cell reaches a voltage of 0.7 V at an average current density of $244\,\text{mAcm}^{-2}$ and an average power density of $170\,\text{mWcm}^{-2}$.

Chapter 4

NICKEL OXIDE FORMATION AT THE ANODE

Nickel oxide decreases the catalytic activity of the anode of a SOFC for the hydrogen oxidation. The formation is partly reversible but has to be avoided to maximize the long term stability at high fuel utilization. The oxidation of nickel could occur due to the reaction with CO, CO_2 and H_2O. If the oxidation of hydrogen and carbon monoxide comply the equilibrium condition including the shift reaction, the oxygen activity can be calculated by one of these reactions.

Different models were supposed for the oxidation of hydrogen at Ni-YSZ (yttria stabilized zirconia) cermet anodes. The differences of the models were found with regard to the location where the chemical and electrochemical reactions occur. Mizusaki *et al.* considered exclusively the Ni surface to be electrochemically active. De Boer *et al.* suggested that interstitial hydrogen and hydroxyl are formed [12]. Bieberle *et al.* as well as Mizusaki *et al.* suggested the presence of adsorbed oxygen on the Ni [11]. Modeling results of the anodic surface coverages of the species O, H, OH and H_2O indicated that the fraction of adsorbed oxygen increases drastically with higher overpotentials [11], which could be an evidence of the formation of NiO at higher overpotentials.

As long as the detailed mechanisms of the hydrogen oxidation and formation of nickel oxide at nickel cermet anodes are not yet clarified, the following assumptions are made for this work. At OCV conditions (Open Circuit Voltage) the formation of NiO can be estimated by the equilibrium activity of the oxygen regarding the H2O-H2 ratio of the gaseous phase. In case of operation and non-equilibrium condition at the anode, the potential of the anode (vs. air) is displaced to less negative values, mainly caused by the activation overpotential. As a result of the following calculation, the concentration overpotential is lower than 10mV. Thus, the concentration-dependent change in oxygen activity is considered to have a minor influence on the formation of NiO. The formation of NiO was observed at negative anodic potentials between -850mV and -650mV (vs. air) [12], which is the same range of oxygen activity governed by the equilibrium of Ni-NiO system. This is why the anodic potential, calculated by activation and concentration vs. air, is used in this work to estimate the formation of NiO.

The formation of nickel oxide is estimated by comparing the anodic potential with the oxygen activity governed by equilibrium of the Ni-NiO system. The equilibrium oxygen activity of $Ni + \frac{1}{2}O_2 \leftrightarrow NiO$ is calculated by [13]

$$\log(a_{O2})_{Ni-NiO} = 8.96 - \frac{24430}{T} \tag{60}$$

where $p_{O2,Ni-NiO} = a_{O2,Ni-NiO} \cdot 1bar$. To avoid the formation of nickel oxide at electrochemical equilibrium condition, it is assumed that the oxygen activity, given by the hydrogen oxidation (Eq.(61)), has to be lower than the oxygen activity of the nickel oxidation (Eq.(60)). The equilibrium of both equations (Eq. (60), Eq (61)) were determined by measurements [13]. The oxygen activity of $H_2 + \frac{1}{2}O_2 \leftrightarrow H_2O$ is given by [13]

$$\log(a_{O2})_{H2-H2O} = 2 \cdot \log\left(\frac{p_{H2O}}{p_{H2}}\right) - \frac{26000}{T} + 5.94 \tag{61}$$

where p_{H2} is the partial pressure of the hydrogen and p_{H2O} the partial pressure of the water. The ratio of these partial pressures is strongly related to the fuel utilization of the SOFC. The results of this empirical approach is similar to those of the thermodynamic equilibrium.

To compare the equilibrium of the Ni-NiO system with the overpotential of an anode, the oxygen activity of the nickel oxidation is converted into a electrical potential related to the potential of a reference air electrode ($p = 1bar$ and $T = T_{SOFC}$). The anode potential vs. air is given by

$$\Delta V_{An/Air} = \Delta V^0_{H2-H2O/Air} + \hat{\eta}_{An} = \Delta V^0_{H2-H2O/Air} + \hat{\eta}_{Act,An} + \hat{\eta}_{Diff,An} \tag{62}$$

where η_{An} is the overpotential of the anode, taking the activation (Act) and diffusion overpotential (Diff) of the hydrogen oxidation into account. The calculation of the activation and concentration overpotential is described in the subsections 0 and 0.

$$\Delta V^0_{Ni-NiO/Air} = -\frac{R \cdot T}{2 \cdot F} \cdot \ln\left(\frac{\sqrt{p_{O2,Air}}}{\sqrt{p_{O2,Ni-NiO}}}\right) \tag{63}$$

The equilibrium potential of the anode related to the reference air electrode can be determined by:

$$\Delta V^0_{H2-H2O/Air} = -\frac{R \cdot T}{2 \cdot F} \cdot \ln\left(\frac{\sqrt{p_{O2,Air}}}{\sqrt{p_{O2,H2-H2O}}}\right)$$
$$\approx -\frac{R \cdot T}{2 \cdot F} \cdot \ln\left(\frac{K_{p,H2-H2O} \cdot p_{H2} \cdot \sqrt{p_{O2,Air}}}{p_{H2O} \cdot \sqrt{p^0}}\right) \tag{64}$$

Figure 13 shows the tendency of the nickel oxidation (grey area), the potential equivalent of the hydrogen oxidation (dashed lines) and the anode potential (full lines) as a function of the temperature and p_{H2}/p_{H2O} ratio.

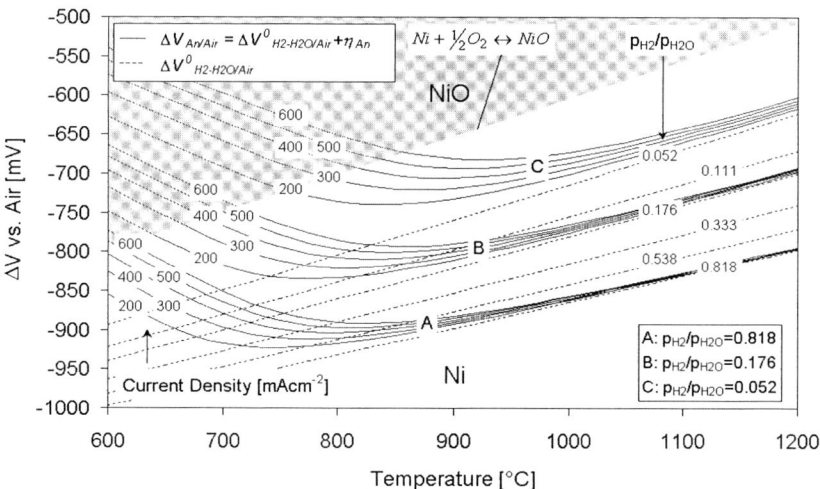

Figure 13. Anode potential (H2-H2O) vs. Air (1bar).

The formation of nickel oxide thermodynamically occurs at oxygen pressures higher than $5.9 \cdot 10^{-11}$ bar (1000°C). This partial pressure is equal to an anode potential of −603mV vs. air (1bar, 1000°C). Guindet et al. observed an influence on nickel oxide formation at an anode potential more positive than –800mV (950°C) [14]. Therefore, the anode potential is chosen lower than –750mV at 1000°C in this study. This value is reached at a p_{H2}/p_{H2O} ratio of 0.11 (fuel utilization = 85%) and at a temperature of 1000°C. This is particularly important at the gas outlet of the anode where high fuel utilizations occur. The influence of the activation and diffusion overpotential on the anode potential is comparably small in this range of temperature and fuel utilization. The activation overpotential increases with lower temperatures and the diffusion overpotential increases mainly with lower p_{H2}/p_{H2O} ratios.

Chapter 5

HIGH FUEL UTILIZING SOFC

The configuration of the high fuel utilizing (*High-uf*) SOFC with a performance of 1000kW is shown in Figure 14. This system consists of two additional SOFC stacks which are connected in series with the first SOFC stack (1a). It is assumed that the air and the pre-reformed fuel (molar fractions: H_2O=0.36, CH_4=0.01, CO_2=0.06, CO=0.07, H_2=0.5) enters the first stack at a temperature of 1073K. The fuel utilization of the first SOFC stack is generally chosen with 85%. The second and third SOFC stack are supplied directly by the cathode gas of the previous SOFC stack, whereas an air bypass for the third stack is necessary to reduce the air flow rate.

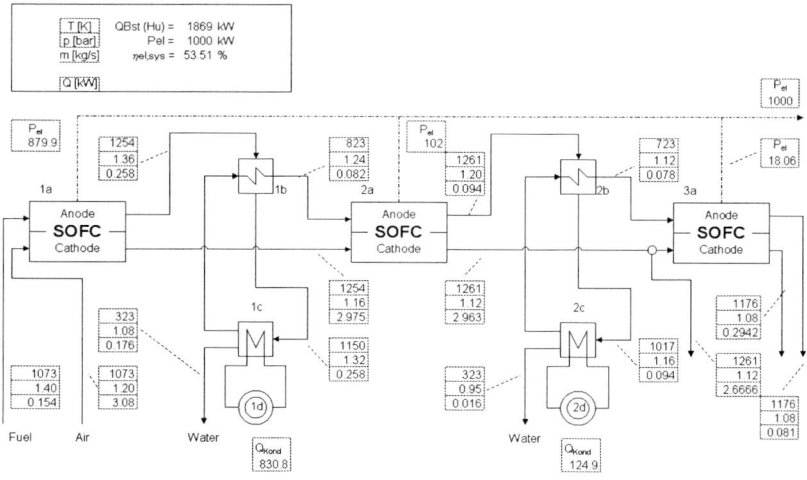

Figure 14. High-uf atmospheric SOFC.

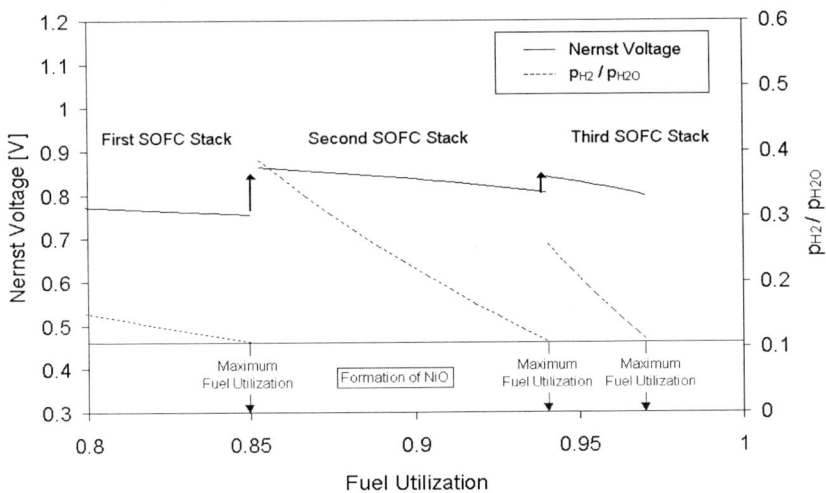

Figure 15. Nernst voltage vs. fuel utilizations.

The anode gas of the first SOFC stack (1a) is led through the heat exchanger (1b) and the condenser (1c) before it enters the second SOFC stack (2a). Thus, the water concentration is decreased and the hydrogen and carbon monoxide concentration is increased. Within the second SOFC stack (2a) the total fuel utilization is increased up to 94% at an average cell voltage of 0.7V. It is assumed that the maximum fuel utilization of each stack is limited by the p_{H2}/p_{H2O} ratio of partial pressures at the outlet of the anode. This limit is chosen at an anode potential of –750mV, whereas a p_{H2}/p_{H2O} ratio of 0.11 is achieved at a fuel utilization of 85% and a Nernst voltage of 0.77V (Figure 15). Within the condenser (1c) the partial pressure of the water is reduced. Thus, the p_{H2}/p_{H2O} ratio is increased up to 0.4 at the condenser outlet (1c). The shift reaction generates water and carbon monoxide by consuming hydrogen and carbon dioxide at the inlet of the second and third SOFC stack. If the shift reaction complies its equilibrium, the p_{H2}/p_{H2O} ratio and the Nernst voltage of the hydrogen oxidation will decrease slightly at the inlet of the second and third SOFC stack. The minimum anode potentials of the first, second and third stack are in a range –760mV and –790mV. Thus, it should be feasible to increase the total further fuel utilization without an amount in the nickel oxide formation. The total fuel utilization of the three stacks increases to 97%. Another benefit of this configuration is an increase in the power density caused by the higher Nernst voltage at the inlet of the second and third stack.

The High-uf SOFC achieves an electrical efficiency of 53% (LHV) at atmospheric system pressure, an average cell voltage of 0.7V and a total fuel utilization of 97%. In this case, the first stack delivers 88%, the second stack 10.2% and the third stack 1.8% of the total power. The efficiency of the High-uf configuration is about 6% point higher than the efficiency of a single stack which operates at a cell voltage of 0.7V and a fuel utilization of 85%.

Depending on the kind of modification, higher system efficiencies could result in smaller system sizes at a constant system power. This applies particularly to the required cell area, which is currently the most imputed item of fuel cell systems. Concerning the High-uf SOFC, a larger cell area has to be installed to achieve a higher efficiency. These opposite influences on the total cell area are investigated in the following consideration. To calculate the size of the additional cell area, which has to be installed for the second and the third stack, the finite difference model (subsection 0) is applied. This model takes the different influences on the local current distribution and local power of the cell into account. For this construction based simulation, the same tubular cell concept of subsection 0 is chosen with different geometric specifications. With regard to the total power of 1000kW, the cell length is specified with 1m and the cell's diameter with 1cm.

Table 4. SOFC input specifications

Cell length [m]	1
Cell inside diameter [m]	$10 \cdot 10^{-3}$
Thickness anode (Ni-YSZ) [m]	$0.05 \cdot 10^{-3}$
Thickness cathode($La_{0.6}Sr_{0.4}MnO_3$)[m]	$0.05 \cdot 10^{-3}$
Thickness electrolyte (YSZ) [m]	$0.1 \cdot 10^{-3}$
Total ohmic resistance of the electrodes and current collectors [Ωcm^2]	0.14
Air preheater tube diameter [m]	$5 \cdot 10^{-3}$

The construction based results of the High-uf SOFC at an average cell voltage of 0.7V and a total fuel utilization of 97% are listed in Table 5. The first stack requires 25,239, the second 3,521 and the third stack 1,064 cells to achieve the respective power of 879.9kW, 102kW and 18.06kW. The average current density of the second and the third stack is lower than the current density of the first stack. This is mainly caused by the lower oxygen partial pressure within the second and third stack and the lower temperature of the third stack. In sum a total number of cells of 29,824 is required. In case of generating a power of 1000kW with a single stack at the same cell voltage and a fuel utilization of 85%, we ought to install 28,683 cells at a system efficiency of 47%. Concerning the High-uf SOFC, the number of cells is merely 4% higher compared to the required number of cells of the single stack at 1000kW. This low amount in the number of cells represents the advantage of the High-uf SOFC, which is caused by the higher system efficiency of 53%.

Table 5. SOFC calculation results

Stack no. (Figure 14)	1a	2a	3a	Total
Stack power [kW]	879.9	102	18.06	1000
Cell voltage [V]	0.7	0.7	0.7	0.7
Total fuel utilization [%]	85%	94%	97%	97%
Number of cells	25,239	3,521	1,064	29,824
Average current density [Am^{-2}]	1554	1292	756	
Cell bundle diameter [m]	2.18	0.99	0.67	

The power density of a SOFC is sensitively influenced by numerous parameters and mechanisms. The cell voltage and the fuel utilization at the design point are two of these parameters. They have to be chosen carefully regarding the economic optimum. Thus, the construction based simulation of the High-uf SOFC is repeated for different cell voltages and fuel utilizations. Furthermore, a single SOFC and a double SOFC configuration are considered to assess the necessity of the additional stacks (2a and 3a, Figure 14).

Figure 16 shows the influence of the fuel utilization and the average cell voltage on the required cell area of the single SOFC (doted lines) and the double SOFC configuration (full lines). The fuel utilization of the single SOFC is varied from 65% to 85%. For the double SOFC configuration it is assumed that the fuel utilization of the second stack is limited by a p_{H2}/p_{H2O} ratio of 0.11 at the outlet of the anode. Thus, a lower fuel utilization within the first stack increases the fuel utilization of the second stack, whereas the total fuel utilization of 94% is nearly constant. Figure 16 shows that the same system efficiency can be achieved by different combinations of the cell voltage and the fuel utilization. Depending on the combination of V_{Cell} and *uf*, the required cell area changes. The average cell voltage is varied from 0.6 V to 0.75 V. The required cell area of the single SOFC configuration can be slightly reduced by decreasing the fuel utilization. Generally, the required cell area increases with higher cell voltages and higher fuel utilizations at a constant power. The single SOFC configuration shows an opposite effect at low fuel utilization (65%) caused by the fixed geometry of the cell design. Thus, the average temperature of the cell decreases and the ohmic losses of the electrolyte increase at lower fuel utilization. With the double SOFC configuration the system efficiency can be increased by 4%-6%-points compared to the maximum efficiency, which is achieved by the single SOFC configuration. The cell area of the first stack can be reduced due to the higher efficiency, whereas the additional cell area has to be installed by the second stack. In sum, the total cell area of the single and double SOFC configuration is nearly constant at equal cell voltages.

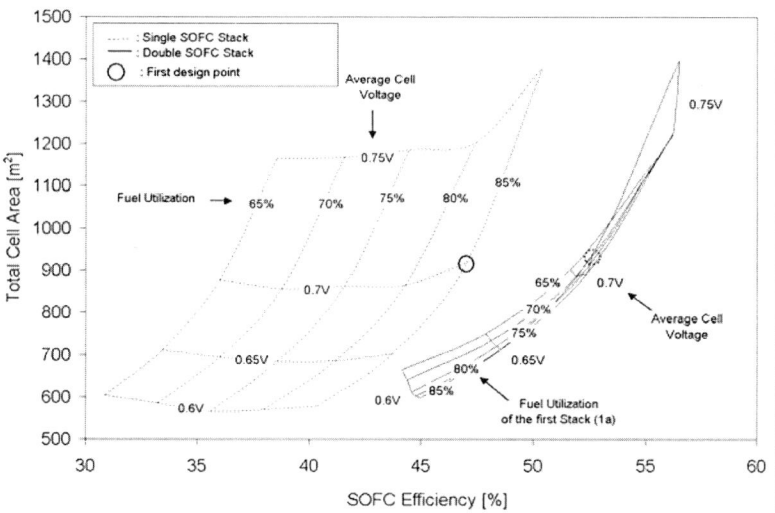

Figure 16. Influence of the average cell voltage and fuel utilization on the required cell area (single and double Stack).

For the triple SOFC configuration (Figure 17) it is assumed that the fuel utilization of the second and the third stack are limited by a p_{H2}/p_{H2O} ratio of 0.11 as well.

Thereby, the total fuel utilization of 97% is nearly constant. With the triple SOFC configuration the system efficiency can be increased by 5%-7%-points compared to the maximum efficiency of the single SOFC configuration. The maximum efficiency of the triple SOFC configuration is about 1%-point higher than the maximum efficiency of the double SOFC configuration. This is caused by the slight amount in the total fuel utilization, which is 94% at the double SOFC and to 97% at the triple SOFC configuration. Thus, it is expected that the triple

SOFC configuration does not give any remarkable advantages compared to the double configuration.

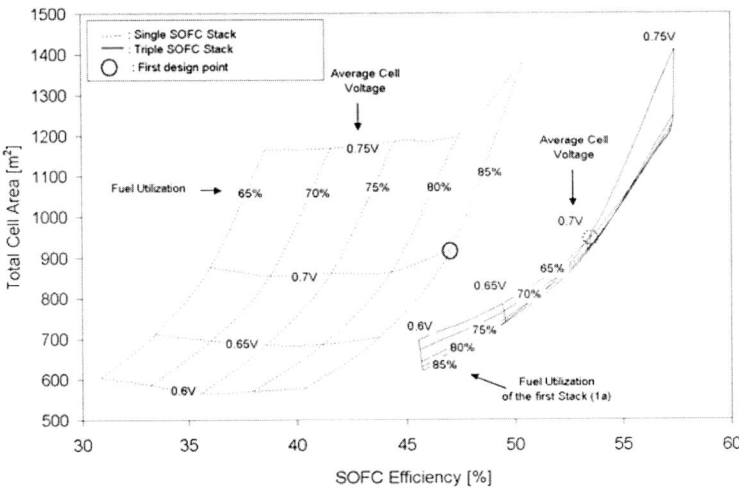

Figure 17. Influence of the average cell voltage and fuel utilization on the required cell area (single and triple Stack).

It can be summarized that the High-uf SOFC configuration significantly increases the system efficiency by using the same cell area as the single SOFC configuration. This could reduce the investment cost at equal system efficiencies or the fuel cost at equal installed cell area. Nevertheless, the potential in cost reduction of pressurized SOFCs, which operates at low excess air in combination with a high efficient turbine, is much higher compared to the High-uf SOFC configuration. Considering the high cooling demand of the condenser, it is expected that the High-uf SOFC configuration is exclusively applicable for stationary application.

The configuration of the High-uf SOFC coupled with a gas turbine is shown in Figure 18. The thermodynamic equilibrium is used to solve the mass balance of the reformer for a predicted outlet temperature. In the reformer the hydrocarbons are converted to H_2, H_2O, CO, CO_2 and residual CH_4 by combining the steam reforming and the shift reaction. The thermodynamic equilibrium of each reaction is given by:

$$\prod x_i^{v_i} = \frac{K_p}{p^{\Delta v}} = \left(\frac{p^0}{p}\right)^{\Delta v} \cdot \exp\left(-\frac{\Delta_r G}{R \cdot T}\right) \tag{65}$$

where x is the molar fraction of reacting species i, p the absolute pressure, T the absolute temperature, $\Delta_r G$ the free reaction enthalpy, $\Delta_r H$ the reaction enthalpy and $\Delta_r S$ the reaction entropy. To avoid thermodynamically the carbon deposition (carbon activity < 1), the Steam / Carbon ratio is chosen with 1.8.

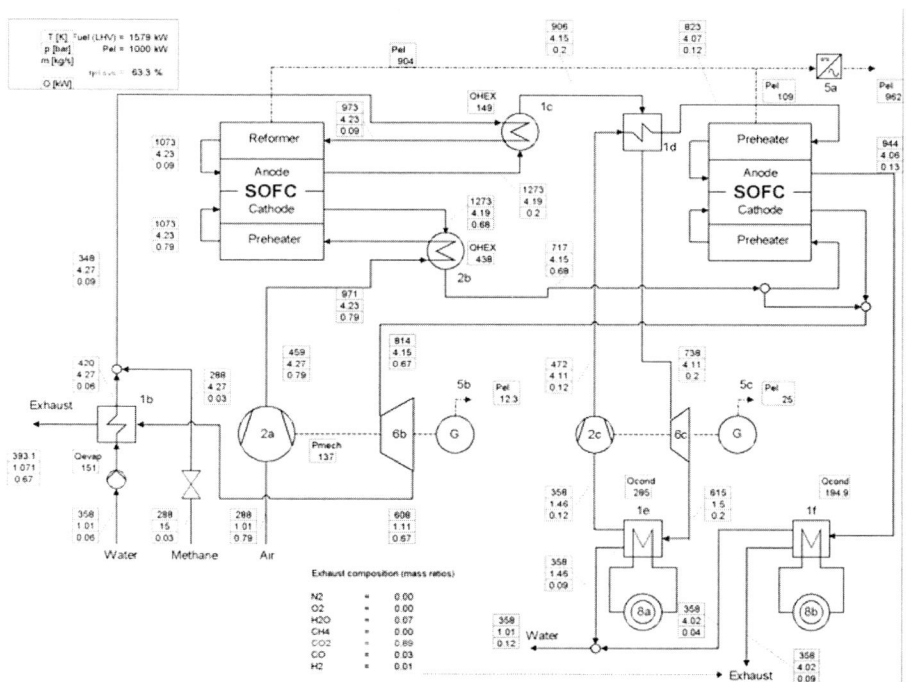

Figure 18. High-uf SOFC-GT.

A constant outlet gas temperature of 1000°C for the first SOFC stack is chosen. To fulfil the energy balance of the SOFC stack, the inlet temperatures of the SOFC stack is calculated iteratively for a total excess air ratio of 1.44. That involves a partial preheating of the reactants inside stack. The second SOFC stack is supplied by the cathode gas of the first SOFC stack since it passed the air preheater 2b. The anode gas of the first SOFC stack is led through the heat exchanger 1d and the condenser 1e before it enters the second SOFC stack. The second gas turbine (2c, 6c, Figure 18) within the anode gas condensing cycle increases the system efficiency by

2.5%-points. Furthermore, the cooling demand of the first condenser 1e (Figure 18) can be reduced by about 40%-points compared to the cooling demand of the condenser of the atmospheric High-uf SOFC [15]. It is assumed that the maximum fuel utilization of the stacks are limited by the ratio of partial pressure p_{H2}/p_{H2O} of 0.11 at the outlet of the anodes. Thereby, the total fuel utilization is increased up to 95%.

With the listed assumptions (Table 6) the High-uf SOFC-GT cycle achieves an electrical efficiency of 63% (LHV) at a system pressure of 4.27bar and an average cell voltage of 0.7V. The system efficiency of the simple SOFC-GT configuration achieves 67% [3]. Thus, 4%-points of system efficiency are taken by the benefit to extract the CO_2 with a minimum of energy.

Table 6. Initial parameters of High-uf SOFC-GT cycle

SOFC average cell voltage [V]	0.7
System Pressure [bar]	4.27
Air ambient conditions (dry air)	288 K, 1.011 bar
Total excess air ratio	1.44
Steam/Carbon ratio	1.8
Isentropic efficiency of compressor/turbine [%]	85
Mechanical efficiency of compressor/turbine [%]	99
Pressure drop SOFC, HEX, Combustor [bar]	0.04; SOFC: 0.01
Thermal losses SOFC, HEX, Combustor [%]	0
Inverter efficiency [%]	95
Genertor efficiency [%]	96

The construction based results of the High-uf SOFC-GT cycle at an average cell voltage of 0.7V and a total fuel utilization of 95% are listed in Table 7. The first stack requires 15,239 and the second stack 2,667 to achieve the respective power of 904kW and 109kW. In sum, a total number of cells of 17,907 is required. Compared to the High-uf SOFC at atmospheric pressure (Table 5), the number of cells can be reduced by 40%-points with the High-uf SOFC-GT cycle. The savings in the number of cells is mainly caused by the pressurization of the stacks and the higher system efficiency, whereas conventional technologies like gas turbines could achieve remarkable advantages with regard to the investment cost of SOFC-GT cycles.

Table 7. SOFC calculation results

Stack (Figure 18)	First Stack	Second Stack	Total
Stack power [kW]	904	109	1013
Cell voltage [V]	0.7	0.7	0.7
Total fuel utilization [%]	85%	95%	95%
Number of cells	15,239	2,667	17,907
Average current density [Am^{-2}]	2645	1821	

Furthermore, the second condenser (1f) increases the carbon dioxide mass fraction to 89%, which could be interesting for CO_2-sequestring applications. High carbon dioxide mass fractions reduce the energy consumption of the CO_2 storage.

Chapter 6

CONCLUSION

The efficiency of non-pressurized SOFC systems mainly depends on the average cell voltage, the electrochemical fuel utilization and the demand of excess air. The fuel utilization is commonly chosen with 85% at Ni-Cermet anodes. The operation at higher fuel utilizations is particularly critical due to the decrement of the Nernst voltage and the formation of nickel oxide at the anode. Both effects are mainly governed by the local hydrogen to water ratio of the anode gas. Therefore, it is essential to understand the local resolution of the gas composition and its influence on the total power density. In this context, analytical and numerical solutions of the integral current density at a constant area specific resistance (ASR) are presented in this study. Different conditions of SOFCs and test procedures require different calculations for the evaluation of the ASR. Three cases at different distributions of the Nernst voltage are considered with regard to the solution of the integral cell area. The results show that the integral determination of the total power density is an essential method to determine the cell's power if high fuel utilization occur along the cell area. It is further shown that the sporadically published estimation of the cell's power density from an average Nernst voltage is at high fuel utilization in deviation to the integral solution.

Based on the numerical solution of the local current density, a novel conceptual solution is proposed which allows improvements in the fuel utilization. The high fuel utilizing atmospheric SOFC configuration is assessed with a special focus on the formation of nickel oxide, system efficiency and the required cell area at a fixed system performance of 1 MW. This system consists of two additional SOFC stacks which are connected in series with the first SOFC stack. The anode gas of the first SOFC stack is led through a heat exchanger and a condenser before it enters the second SOFC stack. Thus, the water concentration is decreased and the hydrogen and carbon monoxide concentration is increased. It is assumed that the maximum fuel utilization of each stack is limited by the p_{H2}/p_{H2O} ratio of partial pressures at the outlet of the anode. This limit is chosen at an anode potential of – 750mV, whereas a p_{H2}/p_{H2O} ratio of 0.11 is achieved at a fuel utilization of 85% and a Nernst voltage of 0.77V. Within the second and the third SOFC stack the total fuel utilization is increased up to 94% and 97%, respectively. The High-uf SOFC achieves an electrical efficiency of 53% (LHV) at an atmospheric system pressure, an average cell voltage of 0.7V and a total fuel utilization of 97%. In this case, the first stack delivers 88%, the second stack 10.2% and the third stack 1.8% of the total power. The efficiency of the High-uf configuration is about 6% point higher than the efficiency of a single stack which operates at a cell voltage of 0.7V and a fuel utilization of 85%. The first stack requires 25,239, the second 3,521 and the third stack 1,064 cells to achieve the respective power of 879.9kW, 102kW and 18.06kW. The average current density of the second and the third stack is lower than the current density of the first stack. This is mainly caused by the lower oxygen partial pressure within the second and third stack and the lower temperature of the third stack. In sum, a total number of cells of 29,824 is required. In case of generating a power of 1000kW with a single stack at the same cell voltage and a fuel utilization of 85%, we ought to install 28,683 cells at a system efficiency of 47%. Concerning the High-uf SOFC, the number of cells is merely 4% higher compared to the required number of cells of the single stack at 1000kW. This low amount in the number of cells represents the advantage of the High-uf SOFC, which is caused by the higher system efficiency of 53%.

The High-uf SOFC gas turbine cycle achieves an electrical efficiency of 63% (LHV) at a system pressure of 4.27bar. The system efficiency of the simple SOFC-GT configuration achieves 67% [3]. Thus, 4%-points of system efficiency is taken by the benefit to extract the CO_2 with a minimum of energy. Compared to the High-uf SOFC at atmospheric pressure (Table 5), the number of cells can be reduced by 40%-points with the High-uf SOFC-GT cycle.

The High-uf SOFC configuration could reduce the investment cost at equal system efficiencies or the fuel cost at equal installed cell area. Considering the high cooling demand of the condenser, it is expected that the High-uf SOFC configuration is exclusively applicable for stationary application. Regarding the high CO_2 concentration at the outlet of the third stack, the High-uf configuration could be interesting for CO_2-sequestring applications.

REFERENCES

[1] W. Winkler. Brennstoffzellenanlagen. Springer Verlag, 2003
[2] P. Nehter; W. Winkler: *System analysis of Fuel Cell APUs for Aircraft applications*. H_2-Expo, Hamburg, 2005
[3] P. Nehter. Thermodynamische und ökonomische Analyse von Kraftwerksprozessen mit Hochtemperatur-Brennstoffzelle SOFC. Shaker Verlag, Dissertation, 2005
[4] Drescher. Kinetik der Methan-Dampf-Reformierung. Dissertation, 1999
[5] A.L.Dicks; K. D. Pointon and A. Swann. The Kinetics of the Methane Steam Reforming Reaction on a Nickel/Zirconia anode. *Proceedings of the 3rd European SOFC Forum*, pp. 249-265, 1998
[6] J. Metzger. *Untersuchung der Stoffumsätze an mit Methan betriebenen Festelektrolyt*-Brennstoffzellen. 1998
[7] M. Mogensen; T. Lindegaard. The kinetics of hydrogen oxidation on a Ni/YSZ SOFC electrode at 1000°C. *Proc. of the 3rd Int. SOFC Symposium*, pp.484-493, 1993
[8] Selimovic. *Modelling of Solid Oxide Fuel Cells*, Applied Analysis of Integrated Systems with Gas Turbines. Dissertation, 2002
[9] S.H. Chan; K.A. Khor; Z.T. Xia. A complete polarization model of a solid oxide fuel cell and ist sensitivity to the change of cell component thickness. *Journal of Power Sources* **93**, pp. 130-140, 2001
[10] E. Achenbach; Ch. Rechenauer. *Dreidimensionale mathematische Modellierung des stationären und instationären Verhaltens oxidkeramischer Hochtemperatur-Brennstoffzellen*. 1993
[11] Bieberle. *The Electrochemistry of Solid Oxide Fuel Cell Anodes: Experiments, Modeling and Simulations*. Dissertation, 2000

[12] de Boer. *SOFC Anode*, Hydrogen oxidation at porous nickel and nickel/yttriastabilised zirconia cermet electrodes. www.matsceng.ohio-state.edu/ims/BDB.pdf, 1998

[13] Rüdiger Dieckmann. Punktfehlordnung, Nichtstöchiometrie und Transporteigenschaften von Oxiden der Übergangsmetalle Kobalt, Eisen und Nickel. Habilitationsschrift, 1983

[14] J. Guindet; C. Roux; A. Hammou. Hydrogen oxidation at the Ni/Zirconia electrode. *Proc. of the 2nd Int. SOFC Symposium*, pp.553-558, 1991

[15] P. Nehter. A high fuel utilizing Solid Oxide Fuel Cell cycle with regard to the formation of nickel oxide and power density. *Journal of Power Sources*, Available online 1 December 2006

INDEX

#

2D, 20, 28

A

activation, 3, 23, 24, 25, 33, 34, 36
activation energy, 23, 24, 25
air, 1, 19, 21, 30, 31, 33, 34, 35, 37, 42, 43, 44, 47
anode, vii, viii, 7, 9, 10, 11, 23, 25, 26, 28, 29, 30, 31, 33, 34, 35, 38, 39, 40, 43, 47, 49
Anode, v, 33, 35, 50
anodes, vii, 1, 22, 25, 33, 44, 47
application, 15, 42, 48
Arrhenius equation, 23
ASR, vii, 1, 7, 8, 13, 14, 16, 17, 47
assumptions, 33, 44
atmospheric pressure, 44, 48
availability, 1

B

barriers, 24
black, 16
Boltzmann constant, 2
bypass, 37

C

capacity, 1, 30
carbon, 2, 21, 29, 30, 33, 38, 43, 44, 45, 48
carbon dioxide, 38, 45
carbon monoxide, 21, 29, 33, 38, 48
catalytic, 1, 23, 33
catalytic activity, 1, 33
cathode, 3, 7, 21, 22, 26, 28, 29, 30, 31, 37, 39, 43
cell, vii, viii, 1, 2, 5, 6, 7, 8, 9, 10, 11, 12, 13, 14, 15, 16, 17, 19, 20, 28, 29, 30, 31, 38, 39, 40, 41, 42, 44, 47, 48, 49
ceramic, 20, 29, 30
CH_4, 22, 30, 37, 42
chemical, vii, 20, 22, 24, 28, 33
chemical reactions, 20, 22, 28
CO_2, viii, 22, 30, 33, 37, 42, 44, 45, 48
commercialization, vii
composition, vii, 8, 22, 47
compositions, 7, 22
concentration, 24, 25, 26, 31, 33, 34, 38, 48
conduction, vii, 30
conductive, 20
conductivity, 1, 20, 28
configuration, viii, 37, 38, 39, 40, 41, 42, 44, 47, 48
Congress, iv
congruence, 17
construction, 39, 40, 44

consumption, 5, 15, 45
controlled, 22
convection, vii, 20, 28
convective, 31
conversion, 5, 22, 23
conversion rate, 23
cooling, 19, 29, 42, 44, 48
cycles, 44

D

degradation, 1, 15
degradation mechanism, 1
demand, viii, 1, 19, 42, 44, 47, 48
density, vii, 1, 2, 3, 5, 7, 9, 10, 11, 13, 14, 16, 17, 20, 22, 23, 24, 25, 28, 29, 30, 31, 38, 39, 40, 45, 47, 50
deposition, 43
deviation, 13, 47
Dicks, 49
diffusion, vii, 1, 3, 8, 22, 24, 25, 26, 27, 34, 36
discretization, 20, 28
distribution, 7, 8, 11, 13, 20, 28, 29, 39
dry, 30, 44
durability, vii, viii, 1

E

economic, 19, 40
electrical, 2, 3, 5, 6, 7, 15, 20, 22, 24, 25, 34, 39, 44, 48
electrochemical, vii, 1, 5, 20, 21, 23, 24, 28, 29, 33, 34, 47
electrochemical reaction, vii, 20, 21, 24, 28, 33
electrodes, 6, 25, 39, 50
electrolyte, 3, 6, 27, 28, 30, 39, 41
electrolytes, 28
electronic, iv, 28
electrons, 6, 27
electrostatic, iv
endothermic, 29
energy, vii, viii, 19, 20, 23, 24, 25, 30, 43, 44, 45, 48

energy consumption, 45
engines, 1
entropy, 2, 7, 11, 22, 25, 28, 43
equilibrium, 2, 21, 22, 23, 24, 33, 34, 35, 38, 42
European, 49
evidence, 33
expert, iv
exponential, 23, 25, 30

F

Fick's law, 26
flow, 2, 6, 8, 10, 11, 21, 22, 27, 28, 29, 30, 37
flow rate, 10, 11, 37
fluid, 22
free energy, 2, 13
fuel, iv, vii, 1, 2, 5, 6, 7, 8, 9, 10, 11, 13, 15, 17, 20, 21, 24, 29, 30, 33, 34, 36, 37, 38, 39, 40, 41, 42, 44, 45, 47, 48, 49, 50
fuel cell, iv, vii, 1, 2, 5, 6, 7, 8, 13, 15, 24, 29, 39, 49

G

gas, vii, viii, 2, 7, 21, 22, 23, 25, 26, 27, 30, 36, 37, 38, 42, 43, 44, 47, 48
gas turbine, viii, 2, 42, 43, 44, 48
Gibbs, 2, 13
Gibbs free energy, 2, 13
graph, 16

H

H_2, 3, 22, 30, 33, 34, 35, 37, 42, 49
heat, vii, 1, 2, 19, 20, 21, 22, 28, 31, 38, 43, 47
heat capacity, 1
heat transfer, vii, 2, 20, 21, 28, 31
high power density, vii
homogenous, 20, 22
hydro, 42
hydrocarbons, 42

hydrogen, vii, 1, 3, 6, 9, 10, 11, 12, 21, 23, 24, 25, 26, 27, 28, 29, 33, 34, 35, 38, 47, 48, 49
hydrogen fraction, 10, 11, 12, 29
hydroxyl, 33

I

injury, iv
interstitial, 33
investment, 15, 42, 44, 48
ionic, 28
ions, 27
isothermal, 8, 14

K

kinetics, 49
Kirchhoff, 20, 24, 28

L

law, vii, 5, 7, 9, 20, 23, 24, 26, 28
lead, 31
leaks, 6
limitation, 1
linear, 9, 10
location, 33
losses, vii, 1, 3, 6, 7, 13, 21, 27, 31, 41, 44
low temperatures, 30
LSM, 25

M

magnetic, iv
matrix, 20, 28
mechanical, iv
methane, vii, viii, 22, 29
models, 13, 33

N

New York, iii, iv

Ni, 1, 2, 22, 25, 30, 33, 34, 39, 47, 49, 50
nickel, vii, viii, 1, 2, 33, 34, 35, 38, 47, 50
nickel oxide, vii, viii, 1, 33, 34, 35, 38, 47, 50
NiO, 33, 34
nitrogen, 26
nodes, 28
non-linear, vii, 7, 10, 11, 13

O

Ohmic, 9, 27, 30
online, 50
oxidation, vii, 1, 20, 21, 23, 24, 25, 28, 29, 33, 34, 35, 38, 49, 50
oxide, iv, vii, 1, 2, 6, 33, 36, 49
oxygen, 3, 6, 10, 26, 33, 34, 35, 39, 48

P

parameter, 5, 23
performance, 1, 8, 13, 14, 15, 24, 37, 47
polarization, vii, 2, 3, 7, 22, 24, 30, 49
pore, 22, 25, 27
pores, 26, 27
porosity, 27
porous, 25, 50
porous materials, 25
power, vii, 1, 2, 5, 9, 10, 11, 12, 13, 14, 15, 20, 32, 38, 39, 40, 44, 45, 47, 48, 50
preparation, iv
pressure, 1, 2, 7, 9, 10, 23, 25, 26, 34, 35, 38, 39, 43, 44, 48
procedures, 8, 47
production, 25
property, iv

R

radiation, vii, 1, 3, 20, 21, 22, 28, 30
radius, 2, 27, 30
range, 9, 11, 16, 34, 36, 38
reactants, 5, 8, 43
reaction rate, 2, 24
reduction, 42

relationship, 23
resistance, vii, 1, 2, 7, 10, 22, 39, 47
resolution, vii, 11, 12, 19, 47

S

savings, 44
sensitivity, vii, 49
series, 37, 47
services, iv
Siemens, 1
simulation, vii, 39, 40
SOFC, v, vii, viii, 1, 2, 8, 19, 20, 28, 31, 33, 34, 37, 38, 39, 40, 41, 42, 43, 44, 45, 47, 48, 49, 50
solid oxide fuel cells, iv, 1
solid surfaces, 21
solutions, vii, 47
species, vii, 5, 7, 22, 23, 24, 26, 27, 33, 43
specific heat, 20
stability, 28, 33
steady state, 28
storage, 45
streams, 21
systems, 1, 19, 39, 47

T

temperature, 2, 7, 8, 10, 13, 16, 20, 28, 29, 30, 31, 35, 36, 37, 39, 41, 42, 43, 48
test procedure, 8, 47
theory, 26
thermal, 1, 20, 28
thermodynamic, vii, 23, 34, 42

thermodynamic equilibrium, 23, 34, 42
thermodynamics, 7
time, 3, 20, 30
transfer, vii, 2, 11, 20, 21, 24, 28, 31
transport, 22
transport processes, 22
tubular, vii, 19, 28, 31, 39

U

uniform, 10
universal gas constant, 2

V

values, 11, 12, 17, 33
variation, 11
velocity, 2, 27

W

water, vii, 3, 9, 10, 25, 26, 27, 34, 38, 47, 48
Westinghouse, 1

Y

YSZ, 25, 28, 30, 33, 39, 49

Z

zirconia, 33, 50